Dear Parent:
Your child's lo

I Can Read Books have introduced children to the joy of reading since 1957. Featuring award-winning authors and illustrators and a fabulous cast of beloved characters, I Can Read Books set the standard for beginning readers. From books your child reads with you to the first books they read alone, there are I Can Read Books for every stage of reading:

SHARED READING
Basic language, word repetition, and whimsical illustrations, ideal for sharing with your emergent reader

BEGINNING READING
Short sentences, familiar words, and simple concepts for children eager to read on their own

READING WITH HELP
Engaging stories, longer sentences, and language play for developing readers

READING ALONE
Complex plots, challenging vocabulary, and high-interest topics for the independent reader

ADVANCED READING
Short paragraphs, chapters, and exciting themes for the perfect bridge to chapter books

Every child learns in a different way and at their own speed. Some read through each level in order. Others go back and forth between levels and read favorite books again and again. You can help your young reader improve and become more confident by encouraging their own interests and abilities.

A lifetime of discovery begins with the magical words, **"I Can Read!"**

The Wildlife Conservation Society and Gorillas

For many years, people thought gorillas were fierce and dangerous. George Schaller, a scientist with the Wildlife Conservation Society (WCS), first studied gorillas in the wild and learned that they were gentle plant-eaters. Later, WCS scientists Bill Weber and Amy Vedder studied both the needs of the gorillas and the people who lived near them, then created a program in Rwanda where visitors paid to see wild gorillas. The money saved the gorillas' forest home from being turned into farmland.

Today, in the eight African countries where gorillas live—Rwanda, the Democratic Republic of Congo, Uganda, Gabon, the Congo Republic, Cameroon, Nigeria, and the Central African Republic—WCS works with governments to conserve parks and with local people to help solve threats gorillas face.

In New York, visitors to Congo Gorilla Forest at WCS's Bronx Zoo have directed millions of dollars to help gorillas in the wild. Dan Wharton, director of WCS's Central Park Zoo, helps lowland gorillas as head of an American Zoo and Aquarium Association's Species Survival Program.

To find out more about WCS and the ways you can help gorillas and other endangered animals, visit www.wcs.org.

With gratitude to Peter Hamilton. Special thanks to Dan Wharton, WCS scientist, director of Central Park Zoo, and expert consultant. Thanks for photographs to Dennis DeMello (front cover, pages 7, 11, 12–13, 14–15, 16, 18–19, 22–23, 29, 31, 32) as well as Amy Vedder and Bill Weber, WCS International Program directors (title page, pages 4–5, 20–21, 24, 28), Michael G. Kaplan (page 9), Doris Friedman (page 10), Stephen Sautner (page 26), and Julie Larson Maher (page 27); photo on page 8: comstock.com.

ISBN 0-439-86668-5

12 11 10 9 8 7 6 5 4 3 2 1   6 7 8 9 10 11/0

Printed in the U.S.A.   23

First Scholastic printing, October 2006

I Can Read!

READING
WITH HELP
2

Amazing GORILLAS!

Written by Sarah L. Thomson

Photographs provided by the
Wildlife Conservation Society

SCHOLASTIC INC.
New York Toronto London Auckland Sydney
Mexico City New Delhi Hong Kong Buenos Aires

WILDLIFE
CONSERVATION
SOCIETY

Gorillas live in families,

like many people do.

A gorilla family is called a group.

In most groups

there are two or three mothers

and their children.

A group may have
one or two young males.
A young male gorilla
is called a blackback.

Many groups also have

one older male.

When a male gorilla

is about twelve years old,

the fur on his back

turns gray or silver.

Then he is called a silverback.

The silverback is often

the father or grandfather

of the babies in his group.

Gorillas are like people

in another important way.

Both gorillas and people
are animals called primates.
(Say it like this: PRY-mate.)
Chimps and monkeys are primates, too.

chimps

Every primate has

a thumb and four fingers

on each hand.

The thumb makes it easier

for a primate

to pick a leaf off a twig

or swing from a branch

or tickle a baby.

All primates take care
of their babies for a long time,
until the babies are ready
to live on their own.
Most often, a gorilla mother
has one baby at a time.

The baby hangs on to the fur
on its mother's stomach.
When it is a little older,
it will ride on her back.

Young gorillas spend hours playing.

They chase each other.

They climb trees

and swing on branches.

One gorilla may stand

on the top of a fallen tree.

The others try to push it off.

Young gorillas learn by playing with other gorillas.

A young gorilla

learns to make grunts

that mean, "Stop! Don't do that!"

It learns to make soft sounds

like a person burping.

These sounds mean, "I'm happy.

Everything is good."

It learns to hit its chest

and slap leaves or tree trunks

and scream.

This means, "I'm very strong!

Watch out!"

A young gorilla

drinks its mother's milk.

Grown-up gorillas

eat leaves and fruit

and sometimes insects or flowers.

They hardly ever eat meat.

Gorillas are big animals.

A silverback can be as heavy

as two grown men.

Gorillas must eat a lot of plants
to feed their large bodies.
They eat for most of the day.

When gorillas are not eating,
they rest.
Sometimes one gorilla
grooms another.

It uses its fingers to pick dirt
or sometimes ticks and lice
out of the other gorilla's hair.
After they rest,
the gorillas walk to a new spot
and eat some more.

When it gets dark,

gorillas build nests on the ground

or in bushes or trees.

They press down leaves and grass
to make a soft pillow.
Every night they build a new nest
in a new spot.

When gorillas grow up,

they often leave their groups.

A female gorilla

may find a new group

or a new male to live with.

A male gorilla lives alone

or with other young males

for a few years.

Then he joins a new family group

or finds some females to live with him.

Some gorillas can live

up to forty years.

There are not many places left

where gorillas can live.

They live in the forests of Africa

where they have food to eat.

But people chop down the trees.

They sell the logs.

They use the land for farms

or dig mines for metal.

People also hunt gorillas.

Many gorillas are in danger.

If people do not help them,

soon there may be none at all.

Gorillas are afraid of people.

When scientists go into the forest

to study gorillas,

they must be quiet and move slowly.

Sometimes they even

chew on leaves like a gorilla

or make gorilla noises.

When gorillas get used to people,

scientists can watch them closely.

They learn where gorillas live

and what they eat.

They watch as young gorillas

grow up and have babies of their own.

Scientists learn what gorillas need
and how we can help them survive.
We can protect forests
where gorillas live.
We can make parks
where gorillas are safe.
We can help people find food
so they will not need
to make forest land into farms
or hunt gorillas to eat.

Gorillas are primates, just like us.

And they need our help to survive.